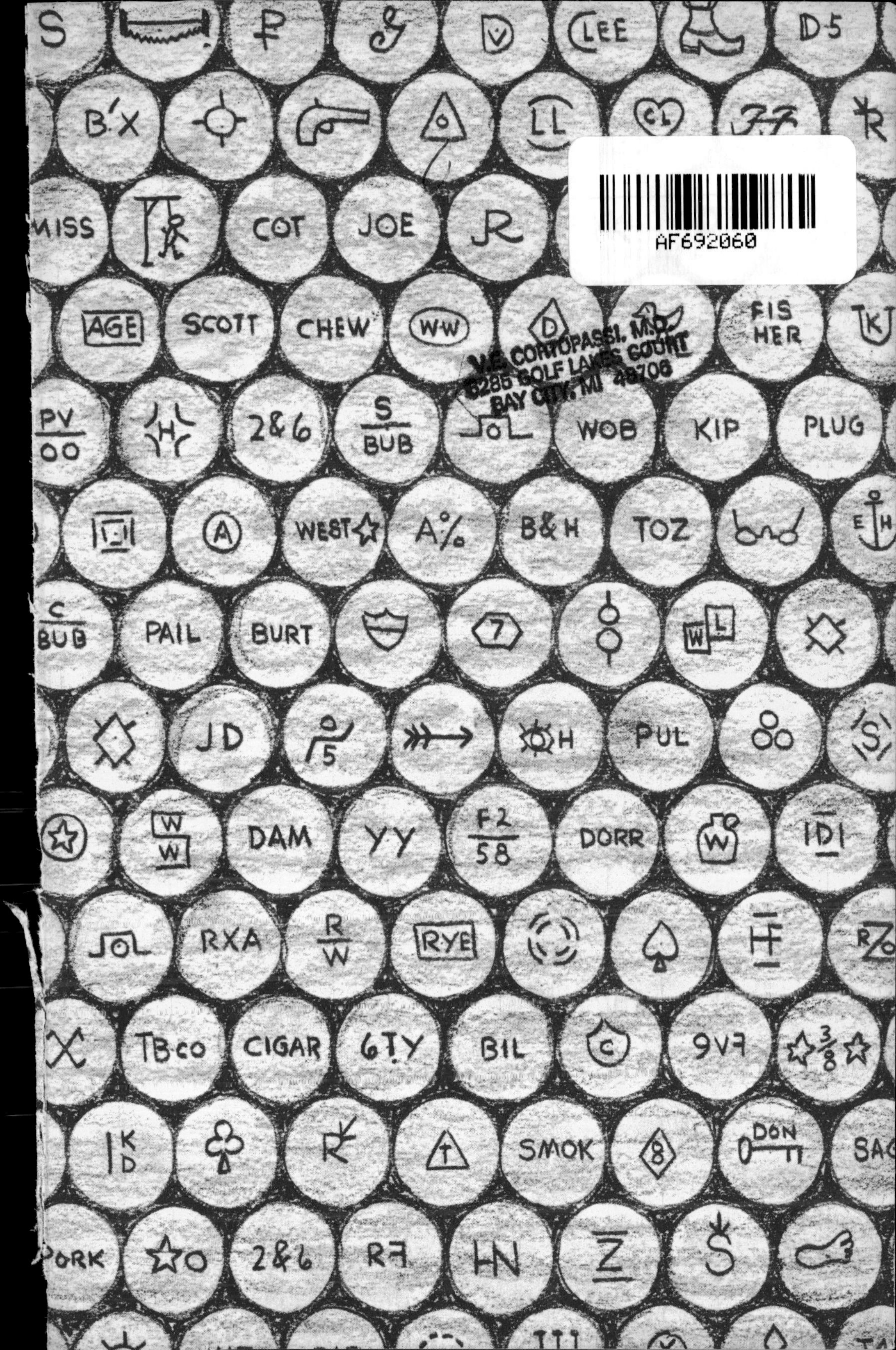
LEE
D-5
MISS
COT
JOE
AGE
SCOTT
CHEW
FIS HER
2&6
S BUB
WOB
KIP
PLUG
WEST
B&H
TOZ
PAIL
BURT
JD
PUL
DAM
YY
F2 58
DORR
RXA
R W
RYE
TB·co
CIGAR
BIL
SMOK
DON
PORK
2&6

The authors of this little book were fortunate in knowing personally and in hearing at first hand the stories, songs and adventures of some of the shanty boys who labored in the Michigan white pine forests and on her rivers and streams. The stories were told by sawyers, fellers, cooks, ink-slingers, sailors, teamsters, swampers, river-hogs, cruisers—men like the MacDonald brothers, Jimmy, Neil and Angus—Ben Sweeney, Frank Cole and Will Foster. These men, like the logging era itself, have passed from the scene forever. To them, and to every other man who played even a brief role during this lusty era—we dedicate this book.

The Story of

LOGGING

the

White Pine

A STORY
WITH
RARE,
OLD-TIME
PICTURES...

COMPLETE
WITH
DICTIONARY
OF LOGGERS'
TERMS

in the

SAGINAW VALLEY

Irene M. Hargreaves *Harold M. Foehl*

Library of Congress Catalog Card Number 64-8151

Printed in the United States of America

Red Keg PRESS

Bay City, Michigan

First printing, September, 1964
Second printing, September, 1965

Book design and layout, illustration selection and text by Irene M. Hargreaves ana Harold M. Foehl

Contents

The bygone era of the vast Saginaw lumber industry survived only a little over four decades—but while it lasted —the white pines and the Valley Boys who cut them played an important role in the building of a still sparsely settled country. This little book does not attempt to tell the whole rough and tumble story—but rather it is intended to give the reader a look at those nearly forgotten times and help him visualize the magnitude and importance of the industry to the Saginaw Valley and to the eastern ports and cities it served.

The white pine logging operations, as described in this book, were repeated many times throughout Michigan—and in all of the Great Lakes States.

POWELL & THORP'S
WESTERN & NORTHERN

Stage & Canal Packet Boat Office,

No. 365 North Market-st. near the Mansion-Houses.

STAGES leave this office every morning and afternoon for Schenectady, Little Falls, Sacket's Harbor, Auburn, Geneva, Canandaigua, Rochester, Lewiston, Buffalo and Niagara—through in Three days to Buffalo.

For CHERRY-VALLEY, COOPERSTOWN, Madison, Manlius, Syracuse and Elbridge to Auburn, every Monday, Wednesday and Friday mornings.

For WHITEHALL, (in connection with the Champlain Steam-Boats,) every Monday, Wednesday and Friday morn- [illegible] , thence to Whitehall by CA- [illegible] *gh in One Day.*—The Steam [illegible] esday and Saturday, at 2 P. [illegible] are in waiting for Montreal. [illegible] OGA SPRINGS every day [illegible] also by w [illegible] t Fort Ed [illegible] g and af [illegible] dy. ☞ [illegible] VELL & [illegible] AN, TH

INDEPENDENT OPPOSITION LINE

FOR

CALIFORNIA

COMPOSED OF STEAMSHIPS
UNCLE SAM, AMERICA & YANKEE BLADE.

The New and Splendid Steamship

AMERICA

1,400 Tons--David C. Mitchell, Commander,

Will sail for Aspinwall, from the foot of North Moore Street, N. R.,

On Thursday, Oct. 20, at 3 P. M.

This Steamer is finished and furnished in a style superior to any in the trade, having a superb Dining Saloon on the upper deck, and will connect with the new and unrivalled Steamship

UNCLE SAM

2,000 Tons—Wm. A. Mills, Commander.

Which will sail from Panama immediately after the arrival of the America's passengers, for San Francisco direct, without stopping at any of the intermediate ports on the Pacific.

The "Uncle Sam" has added to her former honors won on the Atlantic route, by her [illegible] voyage round the Horn, beating all the trips ever before accomplished by six days. It is confidently expected that the passengers by this line will be landed in San Francisco within twenty three days, beating the mails from six to eight days as usual, and ahead of the passengers by any route. The character of these Steamers is so well established, that it would be superfluous to say they are vastly superior to all others on the route, for speed and comfortable accommodations. They are finished and furnished in excellent style, and will be provisioned in the most liberal manner; and as low prices are the order of the day, through tickets will be issued at five per cent. less than the Mammoth Monopoly's [illegible] in their old worn out and slow steamers.

☞ The Isthmus roads are being put in excellent order, and the mule hire has been reduced to $6, as per Panama Star of the 18th inst.

For Freight or Passage, apply ONLY to

E. MILLS, Agent,
51 Cortlandt Street.

MICHIGAN

Saginaw Valley & Lake Shore

LINE OF

STEAMERS.

THE ELEGANT STEAMER

JOHN SHERMAN,

Capt. JOHN STEWART, Master.

Leaves Alpena every Monday, Wednesday and Friday, at 7 o'clock A. M.
Leaves Saginaw every Tuesday, Thursday and Saturday at 9 o'clock A. M. and Bay City on arrival of the Detroit Trains.

THE STEAMER

GEO. H. DUNLAP,

Capt. WM. BROWN, Master.

[illegible] Monday, Wednesday and Friday at 9 o'clock A. M. and Bay City on the arrival of the Detroit Trains.
[illegible] every Tuesday, Thursday and Saturday at 7 o'clock A. M.

Michigania Fever!

Then there's the State of New York,
where some are very rich;
Themselves and a few others have dug
a mighty ditch,*
To render it more easy for us to find
the way,
And sail upon the waters of Michigania—
Yea, Yea, Yea, to Michigania!

In 1848, a fever spread over the eastern states like wildfire. Restless farmers, merchants, teachers, clerks, laborers, speculators, adventurers were packing up their possessions and heading west on canal boats, paddle-wheelers, sloops, in wagon trains and on clipper ships bound for California—and Michigan. It wasn't yellow gold that the speculator sought in Michigan—it was "green gold," which in dollars, was to out-value California's yellow gold by more than a billion dollars![1] Michigan green gold was, of course, White Pine—or the *pinus strobus* —that lofty, graceful, aromatic evergreen that had slumbered for centuries in the vast regions of the upper and lower peninsulas of Michigan. Today—"those trees are gone. They were used to build our civilization."[2]

*The Erie Canal

Twenty-five years before the rush for white pine, the Territory of Michigan was avoided like poison by eastern emigrants and speculators. Settlers heading west skirted the country like it had the plague. Before leaving their farms, villages and towns, to look for "greener pastures," they talked and listened in village meeting houses, in taverns, around the cracker barrel in the country store—and read in newspapers and books about

how the Michigan Territory was inhabited by hostile Indians, covered with "miasmatic marshes and bogs" and infested with mosquitoes—"that are enough to make Philosophy go hang herself, and Patience swear like a Turk or a trooper."[3] Even schoolbooks and atlases showed maps of the lower peninsula overprinted with the words "interminable swamp."[4]

How did the Territory of Michigan earn such a bad reputation? Edward Tiffin, the surveyor general of the northwest can be blamed for much of the unfavorable publicity about the Territory that circulated so freely in the eastern states in the early part of the 1800's. After the war of 1812, Congress set aside six million acres of land in several western states and territories (including Michigan) to be given as compensation to soldiers* of the war—and in 1815, Tiffin was ordered by the land department to begin making surveys of the new Territory. Surveyors began their operations near Detroit, between the Maumee and the River Raisin.[5] They floundered through the marshy, wet forests and after encountering unfriendly Indians, beat a retreat back to "civilization." In his official report, Tiffin wrote, "There would not be more than one acre out of a hundred if there would be one out of a thousand that would, in any case, admit of cultivation," for, "the intermediate space between swamps and lakes, which is probably nearly one half of the country, is, with a very few exceptions, a poor barren, sandy land, on which scarcely any vegetation grows, except very small oaks."[6] Needless to say, all surveys in the Territory were terminated, and Michigan bounty allotments were cancelled.

General Lewis Cass, who was Governor of the Michigan Territory, was incensed by the Tiffin report and, "his loud protest echoed all the way to Washington."[7] He became a compelling, irresistible press agent for the Territory and finally convinced President Madison of its real value. In September, 1816, government land surveys of the Territory of Michigan were resumed.[8] Cass had convinced Washington of the Territory's worth but the damage inflicted by the Tiffin report was very real. It was to be years after Cass negotiated the Saginaw Treaty with the Chippewa Indians in 1819—and after Dewitt Clinton "married the Great Lakes to the Atlantic" at the historic opening of the Erie Canal in 1824, before the easterner seriously considered settling in the vast new Michigan Territory. But, by the early 1830's speculators and settlers slowly began to trickle in.

Michigan became a state in 1837, and by then, in spite of the national economic panic of that year and the acute de-

*Each soldier was to receive 160 acres.

pression that followed, settlers from the east, burning with "Michigania Fever," were streaming into the new state on schooners and paddle-wheel steamers at the rate of two thousand persons a day.[9] The settlers began making clearings in the once silent forests. They felled, girdled and burned great trees, made crude log dwellings and planted patches of corn on land that hadn't seen the sun for a century or more—all the while alternately "shaking their teeth loose with ague chills"[10] and swatting mosquitoes. Somehow, they survived the freezing winters. By 1839 Saginaw's population had swelled to nine hundred and twenty[11] hardy souls and Lower Saginaw, now Bay City, could boast of some twelve to fifteen[12] families. However, no one dreamed that within a few short years the Saginaw Valley would be the first center of commercial lumbering in Michigan.[13] Meanwhile—the eastern speculator was shrewdly buying up Michigan land for future expansion and profitable exploitation.

Lumbermen in New England and in the state of New York soon got wind of the vast white pine stands in the Saginaw River Basin. They eagerly read the Gazetteer[14] and newspapers' glowing accounts of the new state of Michigan. On the Penobscot and Kennebec rivers in Maine, the once immense white pine tracts were melting away under the chopper's axe, and now, everywhere, "dry lumber was in great demand."[15] Eastern vil-

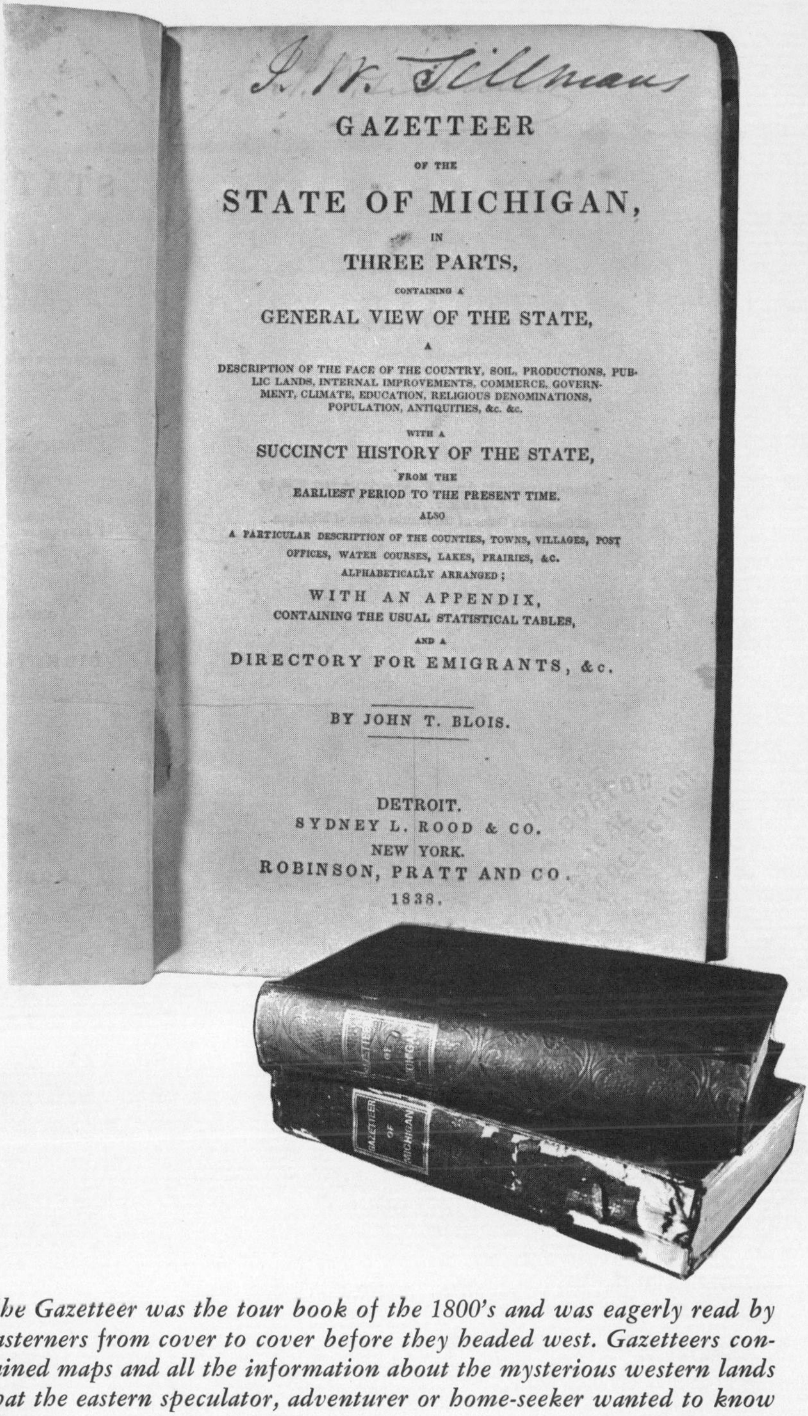

GAZETTEER
OF THE
STATE OF MICHIGAN,
IN
THREE PARTS,
CONTAINING A
GENERAL VIEW OF THE STATE,
A
DESCRIPTION OF THE FACE OF THE COUNTRY, SOIL, PRODUCTIONS, PUBLIC LANDS, INTERNAL IMPROVEMENTS, COMMERCE, GOVERNMENT, CLIMATE, EDUCATION, RELIGIOUS DENOMINATIONS, POPULATION, ANTIQUITIES, &c. &c.
WITH A
SUCCINCT HISTORY OF THE STATE,
FROM THE
EARLIEST PERIOD TO THE PRESENT TIME.
ALSO
A PARTICULAR DESCRIPTION OF THE COUNTIES, TOWNS, VILLAGES, POST OFFICES, WATER COURSES, LAKES, PRAIRIES, &C.
ALPHABETICALLY ARRANGED;
WITH AN APPENDIX,
CONTAINING THE USUAL STATISTICAL TABLES,
AND A
DIRECTORY FOR EMIGRANTS, &c.

BY JOHN T. BLOIS.

DETROIT.
SYDNEY L. ROOD & CO.
NEW YORK.
ROBINSON, PRATT AND CO.
1838.

The Gazetteer was the tour book of the 1800's and was eagerly read by easterners from cover to cover before they headed west. Gazetteers contained maps and all the information about the mysterious western lands that the eastern speculator, adventurer or home-seeker wanted to know about.

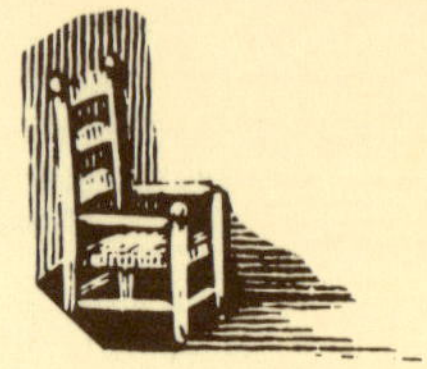

SETTLERS WERE SPREADING OUT OVER THE MAP. THEY

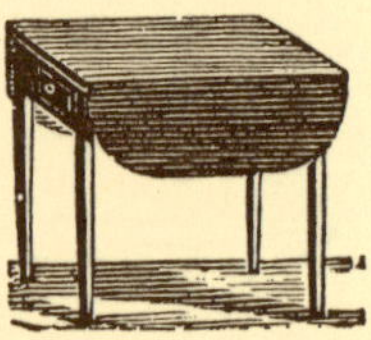

WANTED WHITE PINE FOR HOUSES, BARNS, SHEDS, WAGONS,

FENCES, BUGGIES . . . AND PINE WAS NEEDED FOR BRIDGES,

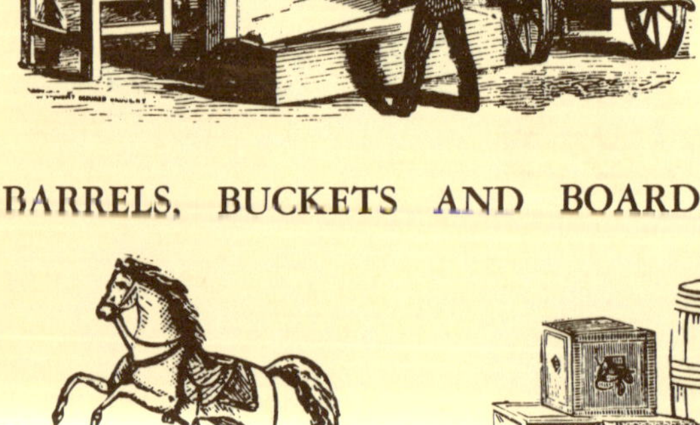

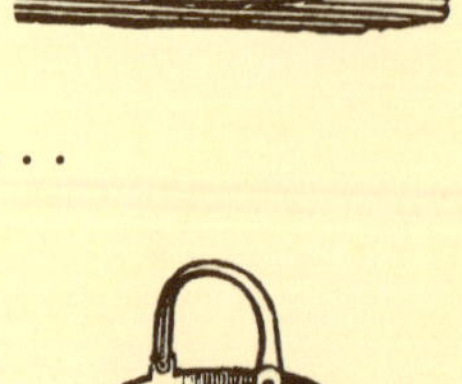

BARRELS, BUCKETS AND BOARDING HOUSES . . .

lages and towns were growing—and settlers were spreading out over the map. They wanted white pine for houses, barns, sheds, wagons, fences. And pine was needed for bridges, boarding houses, saloons, hitching posts, outhouses, sidewalks, brigs and schooners—for steamboats, railroad ties and trestles—and the eastern lumbermen, with their experience and capital, were just the men who would supply it. They set their sights westward and began to send timber "lookers" into the Michigan forests to look for choice lots of pine to cut. The vast yards at Albany, Tonawanda, Buffalo and ports along the Lake Erie shore would soon know about "Saginaw Pine."

The "Looker"

The early timber lookers' estimates of good cutting stands of white pine were crude—but later, when competition became keener, only the most experienced woodsmen were hired to search out the best tracts. The "cruiser" would disappear into the wilds for months on end with a compass in his pocket, a pack of provisions on his back, and armed with government land maps that indicated lakes and streams, types of forest growth, soil conditions and topography. On these maps the cruiser marked the prime timber stands he had carefully selected, estimating closely the number of board feet that could be cut from each section—checking to make certain that the land was close to rivers or streams, because the cut logs had to be floated to the mills. He raced on foot, snowshoe, by dog sled, on horseback or in canoe to register the land at the nearest United

Compass used by Bay City lumberman, John McEwan, during the 1850's.

States land office,[16] usually with competitors hot on his heels. You can be sure he marked out the best stands of pine he could find, for in the later years of the lumbering era, it was customary for a timber looker to receive a quarter of the land purchased for his employers in payment for searching out the rich pine tracts.[17]

The Saginaw River Basin, with its vast network of streams, was especially coveted by eastern lumbermen. The white pine that flourished near its rivers and creeks was tall and straight—with only a slight taper at the top. The big logs had a special buoyancy that the lumbermen liked because they floated like cork,[18] and when they were cut into boards, the wood was almost white, smooth and free of knots. The Saginaw River Valley was a lumberman's dream come true—ready, natural transportation[19] for an abundant crop that he could purchase at prices as low as $1.25[20] per acre—and even lower!

It seems incredible to conservation-conscious people today, that in a brief span of a little over thirty years, all of the thousands of square miles of fine-grained white pine was ruthlessly slashed from the entire Saginaw River Basin—as well as most of the rest of the lower peninsula of the state of Michigan.[21]

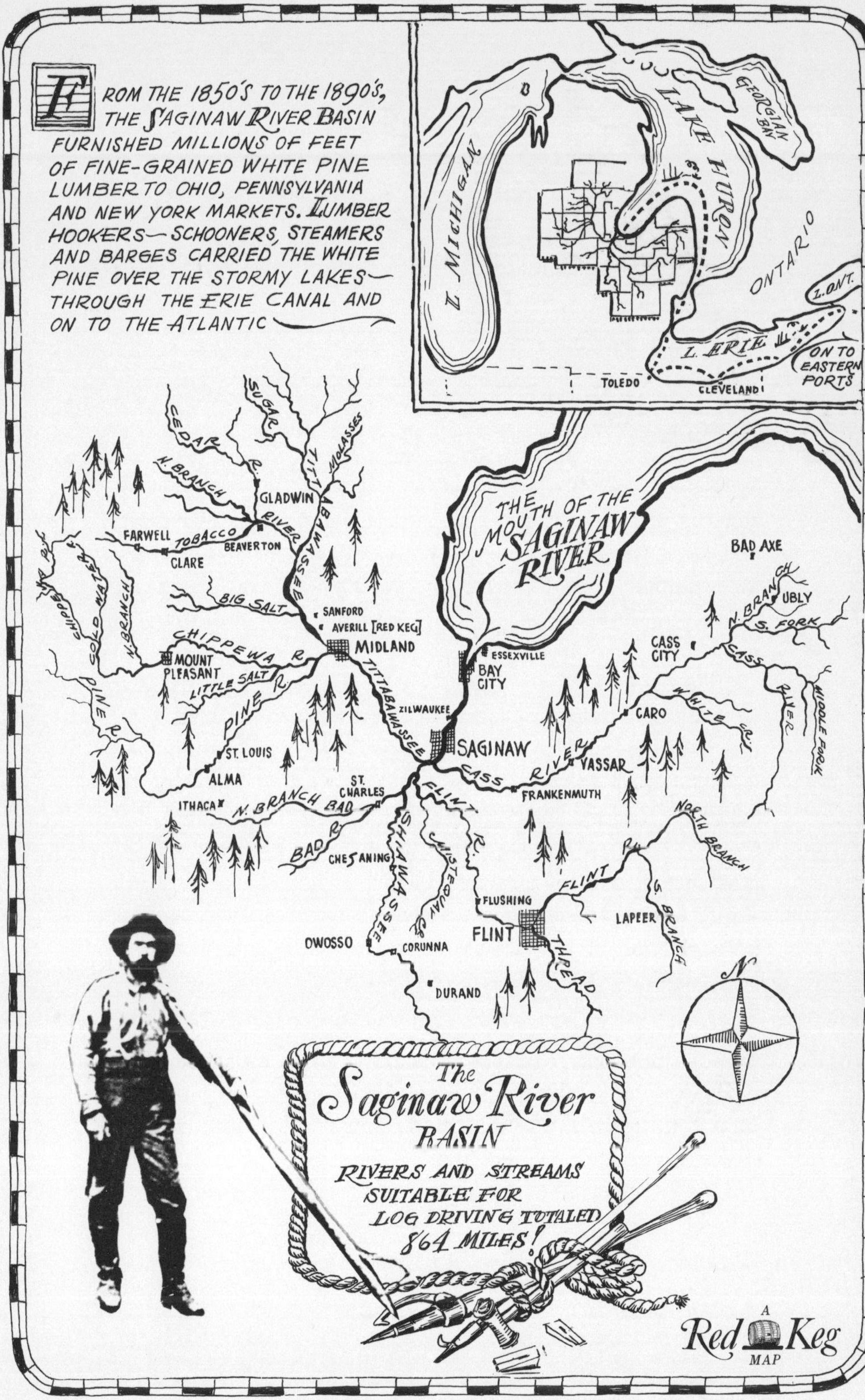
FROM THE 1850'S TO THE 1890'S, THE SAGINAW RIVER BASIN FURNISHED MILLIONS OF FEET OF FINE-GRAINED WHITE PINE LUMBER TO OHIO, PENNSYLVANIA AND NEW YORK MARKETS. LUMBER HOOKERS— SCHOONERS, STEAMERS AND BARGES CARRIED THE WHITE PINE OVER THE STORMY LAKES— THROUGH THE ERIE CANAL AND ON TO THE ATLANTIC
L. MICHIGAN
LAKE HURON
GEORGIAN BAY
ONTARIO
L. ONT.
L. ERIE
ON TO EASTERN PORTS
TOLEDO
CLEVELAND
THE MOUTH OF THE SAGINAW RIVER
SUGAR
CEDAR R.
MOLASSES
TITT
N. BRANCH
GLADWIN
RIVER
TOBACCO
FARWELL
CLARE
BEAVERTON
TITTABAWASSEE
CHIPPEWA R.
COLD WATER
N. BRANCH
BIG SALT
SANFORD
AVERILL [RED KEG]
MIDLAND
CHIPPEWA R.
MOUNT PLEASANT
LITTLE SALT
PINE R.
PINE R.
TITTABAWASSEE
ESSEXVILLE
BAY CITY
ZILWAUKEE
SAGINAW
ST. LOUIS
ALMA
ITHACA
N. BRANCH BAD
ST. CHARLES
BAD R.
CHESANING
SHIAWASSEE
MISTEQUAY CR.
FLINT R.
FLUSHING
FLINT
THREAD
OWOSSO
CORUNNA
DURAND
CASS RIVER
FRANKENMUTH
VASSAR
CARO
WHITE CR.
CASS CITY
CASS RIVER
MIDDLE FORK
BAD AXE
N. BRANCH
UBLY
S. FORK
NORTH BRANCH
FLINT R.
LAPEER
S. BRANCH
N
The Saginaw River BASIN
RIVERS AND STREAMS SUITABLE FOR LOG DRIVING TOTALED 864 MILES!
A Red Keg MAP

Logging camps were crude. Water was dipped from a well or stream— or was hauled in by the barrelful . . . and the rude buildings were dimly lighted by smoking candles or kerosene lamps.

The Camps

After the appearance of the timber cruiser, it wasn't long before the Saginaw River Valley forests and streams were echoing with the ring of the chopper's axe, the rasping of the cross-cut saw and the shouts and curses of the shanty boys and riverhogs. Lumbermen were scrambling to invest, "to cut and get out."[22]

In the autumn, crude logging camps were hastily hacked out of the dense woods, and all winter long trees were felled, "bucked" into logs, and decked high on miles of rollways along frozen river banks, where they awaited the spring "breakup" and the big log drive to the mills.

The logging camps in Michigan were more "refined" than the camps in Maine—but even so, the life of the Michigan shanty boy who labored in them was no bed of roses. He worked long hours and earned between $25 to $30[23–24] a month, plus board. Before the introduction of logging trains and year 'round logging, he began his employment in the woods in October or November, before the snow fell. But earlier, before the full camp crew was hired—"road monkeys" cut and slashed a tote road, leading into the forest where the camp buildings were to be erected—and over this uneven road the tote teamster hauled camp supplies and equipment by the wagon load. Hay, grain, beans, molasses, tea, tobacco—barrels of salt pork and "red horse" (salted beef), lard, prunes, potatoes, flour, dried apples—logging tools, blacksmith's equipment, blankets, candles, kerosene, kitchen stoves and pots and pans—to name just a few necessary items needed to set up the camp. While supplies were being hauled in—the "wood butcher" or camp carpenter threw together primitive camp buildings that would be home and village to the loggers from the first freezeup until the spring thaw.

Not all logging camps were alike. They ranged from "haywire" camps—fly-by-night outfits, "strung together with

The tote teamster hauled all supplies, tools and logging equipment into camp by the wagon or sleigh load before the full camp crew was hired.

haywire"—to the bigger, better managed outfits. A typical logging camp might consist of bunkhouse, a combination cook-shack-dining room, office-living quarters for the bosses, a barn for the horses or oxen—and if the men were lucky, a milk cow. There was a blacksmith shop where camp equipment and tools were made and repaired, saws sharpened, and where horses and oxen were shod. Sometimes a company store, or "van," where the boys could buy a pair of mitts, woolen socks, tobacco or "Hinkleys Bone Liniment for Man or Beast—" was included in the camp setup.

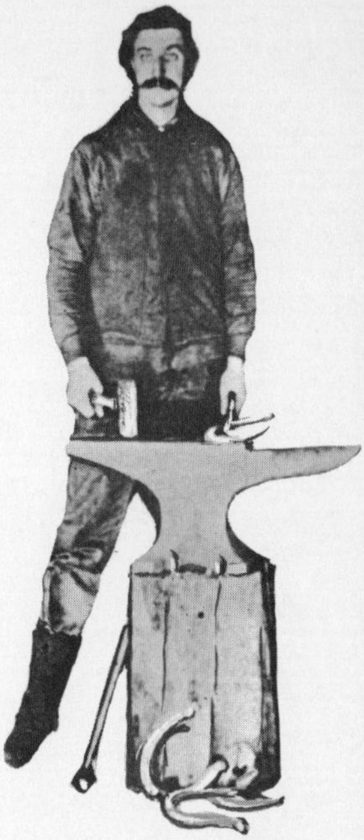

The "Iron Burner."

The bunkhouse was "home" to the shanty-boys. It was a low, squat building, built of massive logs—the walls chinked with clay and swamp grass, and the roof made of wooden "shakes" or tarpaper. Inside, double bunks lined two or three of the walls and a long bench, called the "deacon's seat" was placed against the free wall. The bunks did not have springs or mattresses—the men threw their blankets over spruce or hemlock boughs to soften the rough boards—and for pillows, they rested their heads on their "turkeys," flour or grainsacks filled with their personal belongings. There were usually no windows in the shanty—but then, who needed them? The men were in the woods before dawn and didn't return to the bunkhouse until after the sun went down. In the 1850's and early '60's, a "camboose" or fireplace was built in the center of the bunkhouse—the smoke rising (sometimes) through a rude chimney made of sticks and clay.

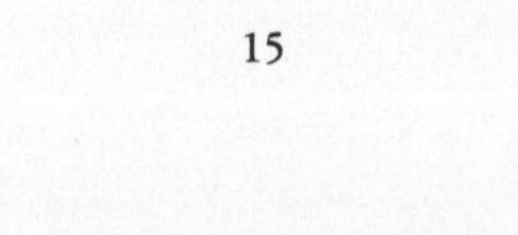

"Cookee" blowing wild trills and arpeggios on his tin horn or "gabriel," calling the shanty boys to a hearty meal in the chuck shack. Man and wife teams were often hired by logging companies to cook for the camp.

In later years, an iron stove stood in the center of the room. It heated and lighted the bunkhouse and was where the shanty boys dried their wet socks, shoepacs and mitts, hanging them up to dry on poles or haywire that was strung up around the stove.

The day's labor in the camp began early. The first ones to pile out of bed in the morning were the cook and his helper, the "cookee." They were up at 3:30 A.M., while stars still shone in the bitter cold night. While cookee started the fires in the kitchen, bunkhouse and bosses' shanty—the cook was already banging pans on the big iron cooking range in the kitchen. At 4.00 A.M. the cookee called the teamsters who fed and harnessed their teams—and at 4:30, the rest of the crew rolled out of warm bunks. Fifteen minutes later—still before dawn—cookee blew a loud signal on his long tin horn—or beat a tattoo on the "gut hammer," and they filed into the cook shack and "fell to" over a breakfast of fried potatoes, sowbelly, pickled beef, sour-dough flapjacks, molasses syrup, gravy, cookies (the size of stove lids), doughnuts and steaming cups of boiled tea or coffee that was guaranteed to make weak men strong—or vice-versa. At noon, to save precious daylight, a meal called "flaggins" was brought out to the men, loaded on a sled or wagon.

After a long day of savage, back-breaking labor in the woods the tired shanty boy trudged back to the bunkhouse and waited for the call to supper. He ate another hearty meal in silence (an unwritten law in the lumber camps), got up from the table and went back to the bunkhouse where he might write an ungrammatical letter (the shanty boy wasn't apt to have much schooling) to his sweetheart or the folks back home—mend his socks, sharpen his tools or play a game of "smear"—

A well fed camp crew was usually a happy one, so the cook was an important person in camp—and he reigned supreme in the cook shack. He did all the cooking and his helper or "cookee" peeled potatoes, washed dishes, chopped firewood, made fires and did a thousand other menial tasks. Garbage disposal was no problem—most camps had a pig penned near the back door to take care of the leftovers —if there were any.

Cookee's tin gabriel sometimes reached astonishing lengths—some were as long as eight feet!

all by the flickering yellow light of a candle, a smoking kerosene lamp or the glow of the red hot stove. He smoked one last pipe —and by nine o'clock, he and every other manjack was in his bunk, exhausted by the long day's labor.

The men went to sleep at night in an over-heated, smoke-filled, steamy room. The only sound that could be heard was the snap-cracking of the dying fire—the sizzle of rows of wet socks and shoepacs steaming and drying around the stove—and the snores in various keys, dischords and "harmonies," coming from the depths of each bunk. The cherry-red glow and warmth of the stove soon disappeared into the darkness and the icy wind whistled through the chinks in the walls. But the boys slept, "like logs," until they were rudely awakened before dawn the next morning.

Shanty boys worked hard all week long and Saturday night was their night to howl. If the logging camps were close enough to town or village saloons, the men drove into town in the tote

wagon or sleigh for some excitement. If the camp was in a remote section of the country, the shanty boys entertained themselves. There was usually a fiddle player, harmonica or dulcimer "artist" in camp—or one who could recite verses, spin a yarn or sing a song. He was known as the "deacon," and his repertoire might include "The Timekeeper's Lament," "Six Whistles," and "Sixteen Men in a Pine-Slab Bunk." When the boys roared out in a chorus,

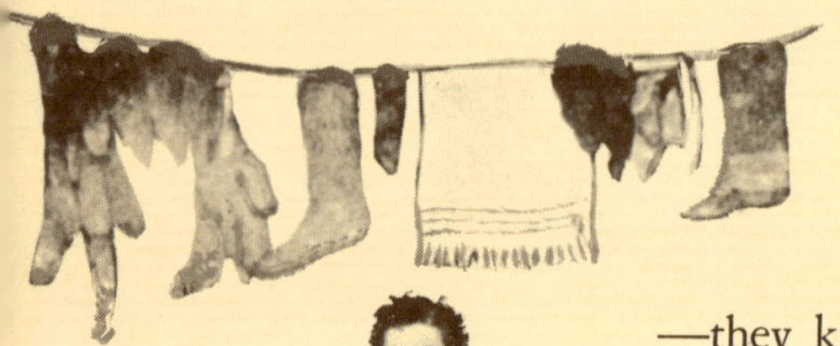

"Beans are on the table;
Daylight's in the swamp,
Hey, you lazy Shanty Boys,
Ain't you ever gettin' up?"

—they knew it was "their song"—and nobody elses! When the fiddle player struck a loud chord for a "hoe down"— the men grabbed their partners* and stomped the hard dirt floor or pine boards in time to the music until the roof nearly caved in.

Sunday was usually "boil up" day, when the shanty boys deloused their blankets and clothing by boiling them in a deadly mixture of scalding water, yellow laundry soap and Peerless tobacco. The losing battle between the boys and the grayback was traditional and perpetual—the loser always being the shanty boy. "Sky pilots" or travelling preachers sometimes rode into camp by sled and held services for those who would listen. Sunday was a day of rest, gabbing and waiting for Monday. So it went through the long winter months of cold, wet, dangerous, fatiguing labor of felling trees and getting them to the rollways.

*The "ladies" wore a scarf or bandana tied around one of their arms.

Sunday was the only day the shanty boys rested. They cleaned up, sometimes shaved, visited, played cards—and, occasionally posed for their pictures for the itinerant commercial photographer.

The shanty boys needed and deserved a rest on Sunday—the rest of the week they were in the woods before dawn—and did not return to the camp each day until after darkness had fallen.

Before the "fellers" went to work cutting down the pines—a gang of "swampers" cleared the cutting ground of brush, making an open space for the cut trees to fall upon.

Before the days of the "cross-cut" saw, the "choppers" laboriously felled the trees by axe alone.

To cut through the bole of the pine, one sawyer worked at each end of the big cross-cut. Wooden wedges were driven into the cut as sawing progressed—thus, the tree fell in the direction the feller had determined it should go.

At the cry of "Timberrrr!" men scurried out of the way of the "widow makers." The big tree would tremble, sway slightly, then go crashing down to the earth. In the above picture, sawyers are squaring the butt end of a fallen pine.

After the pine lay on the ground, the swampers cut off all of its branches. Sometimes the whole tree was dragged to another place for cutting up. The "Canadian way" was to cut it where it was felled— "buckers" cutting the tree into 12, 14, 16, 18 or 20 foot length saw logs. The ends of the logs were "rossed" (bark peeled off the ends) so they would skid more easily over the rough logging trail.

Travois, skidding sleds or "go-devils," were common ways to move the logs to the skidway or "cross-haul."

Logs were decked or stacked into high piles at the skidway. Highly skilled men called "top loaders" worked on top of the decked logs with cant hooks. One slip and the whole pile of heavy logs would go crashing down—perhaps killing or seriously injuring a man or two.

Giant-size sleighs were used to haul the logs from the skidways to the rollway, located along the frozen river banks. But in the summer—and when there was little or no snow in the winter months, "Big Wheels," pulled by teams of horses transported the logs to the rollway. In 1873, Scott Gerrish introduced the steam locomotive and narrow-gauge railroad to the logging industry (near Farwell, Michigan), thus permitting companies to log all year around.

Big Wheels. Standard size wheels were ten feet in diameter, high enough to clear stumps—and strong enough to carry logs from twelve to one hundred feet in length.

Although railroads came into use to transport logs, they never displaced the horse-drawn sleigh or the rivers in the Saginaw River Valley. Sleighs were loaded with logs at the skidways. Men with cant hooks guided the logs as they were rolled up the skids by means of a chain hitched to a horse. The horse was carefully driven ahead while cant-hook men guided the log up the pile on the sleigh. A loaded sled of sixty thousand pounds was not an unusual one.

A stout team of horses had no trouble pulling a sleigh as heavily laden as the one shown here. The sled's heavy runners traveled over sleigh roads slick with ice—the horses, shod with sharp, caulked shoes, could get good footing for an even pull on the heavy load. "Snubbers" were used to brake a loaded sleigh going down hill.

The "sprinkler" was driven over sleigh roads at night, dribbling water from the water-tight tank, over the road surface—where it froze instantly.

The rollway or banking grounds. This is where the shanty boys' job ended and where the river-drive started in the springtime—when the ice melted and the driving streams ran high.

Hundreds of logging camps all over the Valley were engaged in the same kinds of activity. Besides the exhausting labor in the woods—records —time and tallies were kept, thousands of meals cooked, lamps filled, fires lighted, wood chopped, dishes washed, stock fed, clothes dried out, harness and machinery repaired, saws and axes sharpened, supplies toted—never ending routine and hard work for the logger until the soft winds of April began to break up the ice on streams and rivers. Then it was time for the shanty boys to change shoepacs and rubbers for "corked" boots. It was time for the river drives—when the logs on the rollways were broken out and driven by the "whitewater men" down the hundreds of miles of streams to the sorting booms at the booming grounds—and then on to the sawmills.

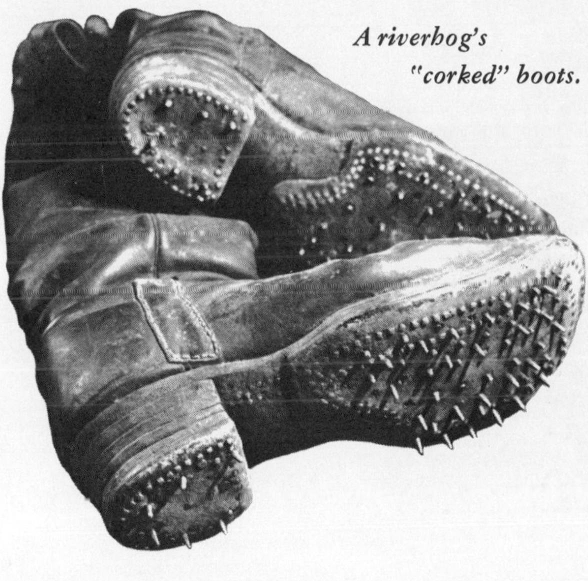

A riverhog's "corked" boots.

Riverhogs.

River Drive

Transporting the high-decked logs from the rollways to the sawmill was often the most costly,[25] and the biggest headache of all lumbering operations. At the first signs of a spring thaw, log dams holding back the waters of the rivers and creeks were checked for "breaking out." The big day would arrive—the water in the streams had risen to just the right "driving pitch." Dams were broken and the key logs in the rollways pulled. Riverhogs, with ready pike poles and peaveys, cautiously watched the decked logs spill out with a deafening roar—cascading into the boiling jets and streams of released water. There they furiously attacked the spinning, twisting logs to keep them floating freely down the stream. The river drive was on.

The finger-like tributaries of the Saginaw—the rivers Tobacco, Salt, Molasses, Cass, Tittabawassee, Bad, Shiawassee, Pine, Flint, Cedar, Chippewa—and *their* tributaries and branches, were soon choked with logs, riverhogs, equipment, cooks, clerks and wanigans. The confusion of thousands of logs dumped into the streams from the many logging camps—all massed together and floating down rivers and creeks—was like a nightmare to the greenhorn. But to the alert, sharp-eyed sorters at the booming grounds, whose job it was to separate company logs, (bearing the same marks), into individual pocket booms,—the sorting was a comparatively simple job.

Riverhogs lived dangerously, but made more money than the shanty-boys. Their pay was about $3.50 a day (plus board).

Wanigans followed the river-drive down the streams and served the drive as kitchen, supply room and office. This photograph shows cooks and helpers who prepared food to sustain the hungry riverhogs. Man standing in dugout is the waterboy, whose job it was to supply the wanigans with fresh water and other supplies.

Throughout the long winter months, as the pine trees were felled, sawed into lengths and dragged by teams to the skidways —the scaler, with his "cheat stick," or loggers' rule, measured and recorded the number of board feet* in each log. Then, with heavy marking hammer, he struck both ends of the log with his company's mark. The mark's design, raised from the

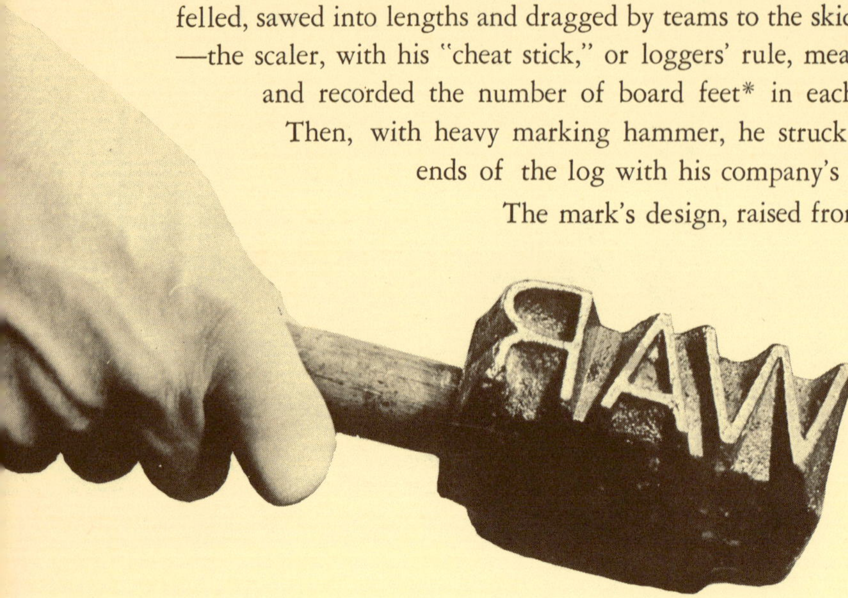

hammer's striking surface, made a distinct brand that was easily visible to the sorters at the booming grounds. The log mark was to the logger what the cattle brand was to the cattleman in the west—protection of ownership. These marks became symbols of law and order in the logging industry.[26]

But even with the enforcement of log mark registration, log piracy was a common practice on the rivers in the early years of logging. One way to steal logs was to wait for the

*The scaler figured a foot of lumber to be a piece 12 inches square and one inch thick.

spring drives . . . the thieves would hide along the streams and, as the logs floated past, they would swiftly snake out their long pike poles and gaff a few valuable prizes. The ends of the stolen logs were sawed off by the pirates—or the marks were skillfully obliterated, and the original logging company's mark was replaced by the thieves' mark.

Pike pole.

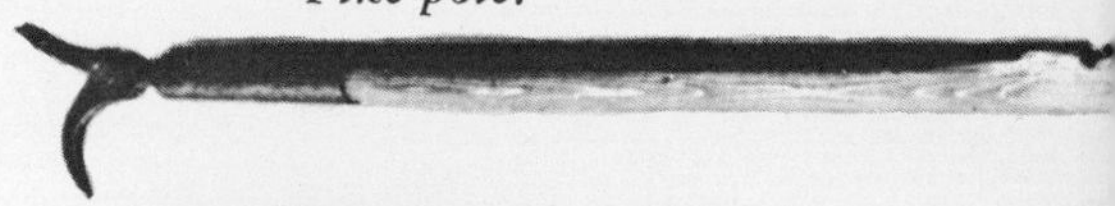

UNIDENTIFIED

WELLS, STONE & CO.

EDDY, AVERY & EDDY

UNIDENTIFIED

MELCHER & NERRETER

MIDLAND SALT & LUMBER CO.

Sawed off ends of stolen logs with owners "marks" still clearly visible on them. They are from the collection of Louis Groulx of Essexville, Michigan, who found them along the Saginaw River.

Stealing logs was a risky business because the riverhogs could always be seen working on the river and along the river banks, releasing, with their long pike poles, the stray logs hung up on islands, snags or sandbars—or they would be riding the logs down the streams, looking for trouble, whether it was in the form of log stealers or log jams.

One of the most infamous timber thieves was a legendary character called LeCoq. His operating area was the Cass river—and you can bet the alert riverhogs and the law were usually looking out for him. He was known throughout the logging camps as a clever thief and many a bunkhouse "deacon" entertained the shanty boys by spinning tall yarns about him on long winter evenings. There is one story about LeCoq that he probably never lived down. It seems that one spring he and another man each made up a raft of choice pine logs, then made their way down the Cass, heading for a Saginaw mill. It was dark and windy by the time they floated into Saginaw, so they were glad to tie their rafts to the dock piles and find a boarding house where they could spend the night. When LeCoq and his companion were in bed, LeCoq lay still, pretending to be asleep, and waited for the even snores that eventually came from his friend. Then he stealthily crept down the stairs and made his way to the dock where they had left the log rafts. He untied the ropes of one of the rafts (supposedly his companion's) and silently poled it to the mill. He collected the money for the logs, then returned to the boarding house, and to bed. The next morning, LeCoq innocently went with his friend to the dock and much to LeCoq's consternation, he found his own raft gone and his companion's raft still tied to the piling. The only explanation was that the wind had moved the logs, and in his haste to steal the raft—and hampered by darkness—LeCoq, the clever thief, had stolen his own logs!

LeCoq wasn't the only thief in the logging business. The sawmill and logging companies were always on the alert to make an extra dollar, and it was not unusual for the mills and loggers to steal from each other. It was a greedy business—one that required all of the protection of ownership that the individual log mark could afford.

As the years passed, county records of registered log marks were either lost or destroyed, so it is difficult to estimate the number of mark designs that were used in the Saginaw Valley. But one would be safe in saying that thousands of log marks were registered through the years.[27] Not only did the logging companies strike their marks on the logs they cut, but their creditors had the right to send their men into the woods and put their brands on the logs too! Each logging camp had its own log mark design and each year the marks were changed[28]—so it is no wonder that the man not accustomed to logging operations would shake his head in wonder at the seemingly impossible task of identifying and sorting the millions of feet of logs that were cut and driven to the mills during the hey-day of the logging industry. The Tittabawassee River Boom Company, that operated at the mouth of the Tittabawassee River at its intersection with the Saginaw, sorted and rafted over ten billion feet of logs between the years of 1864 and 1894[29] —and this figure represents the work of just one booming operation!

WHITNEY & REMICK

MELCHER & NERRETER

BLISS & VAN AUKEN

SEWELL AVERY

BRIGGS & COOPER

W. R. BURT

C. S. BLISS CO.

GEBHART & ESTABROOK

BLISS & TYLER

T. E. DORR & CO.

JAMES DAVIDSON

C. K. EDDY & SON

WRIGHT & KETCHUM

MARTHA HAY

PALMERTON
WOODEN-WARE CO.

C. H. PLUMMER

NESTOR & HOYT

WALL & WEBBER

JOHN F. RUST

WYLIE BROS.

H. W. SAGE CO.

MUMFORD & AVERY

R. S. SOVEREIGN

F. H. HALL

Log marks were many and varied. Shown is a sample of marks representing names that are well known in the Saginaw Valley, even today.

Sorting pens at the booming grounds—showing saw logs passing down the central channel of the boom. Sorters can be seen poling company logs, bearing the same log marks, into their individual pens with their long pike poles. The sorted logs were then made into long "stringers" (rafts) and towed to the sawmills on the Saginaw River.

Log rafts were held together with rope fastened to the center of each soft pine log by a hardwood rafting pin.

Sawmill Towns

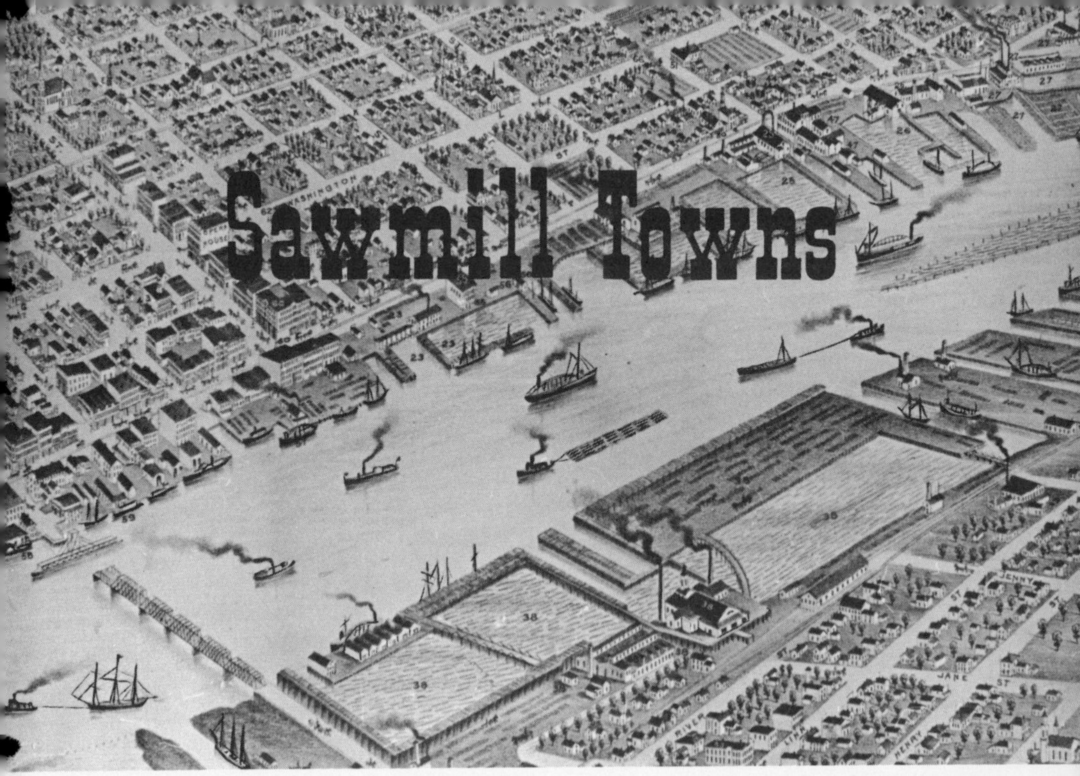

1869—The Saginaw River at the Third Street Bridge, Bay City. Henry W. Sage mill is in the foreground. Note tug pulling "stringer" of logs.

Saginaw and Bay City were the sawmill towns of the Valley[30] —one town near each end of the twenty-mile stretch of the historic Saginaw River. By 1854 there were twenty-nine sawmills and nine more being constructed on the river[31]—and by the 1870's, there were over one hundred mills[32] running full blast, night and day, between Green Point, near Saginaw—and the river's mouth below Bay City. Mill hands worked twelve-hour shifts, six days a week, operating the saws that chewed up the logs after they were towed in great rafts from the booming grounds to the mill ponds.

Millhands labored six days a week on twelve hour shifts. In 1874, the average daily wage was $1.75—but in 1877 it was dropped to $1.25.[33] *Note barrels of water on roof ready to be tipped at the cry of "fire!" Logs were hauled up from the mill pond in an endless procession, into the mill by the "bull chain." Above—Lumber as far as the eye can see. The E. B. Foss & Company dock and yard No. 1, Bay City.*

The mill was "one of the sights of the valley; a monster of the woods, with more sets of teeth than a wholesale dentist, and with an appetite for forests which no purveyor can satisfy, and which puts all calculations astray."[34] Logs were sawed into millions of board feet of lumber, which were stacked in towering piles on river-front docks to dry and await loading on the shallow-draft lumber "hookers"—the lake schooners, barges, and steamers that sailed daily out of the river mouth, bound for the eastern markets.[35] Drummers hawked Saginaw pine throughout Indiana, Ohio and Pennsylvania, bumping over corduroy roads and mud holes in carriages and wagons—vying with one another to take orders for their company's product.

Each year brought changes in mill saws. Shown is the "circular" which came into use around 1850. After 1876, the band saw with its thinner "kerf" gained first place in the mills. Then the "gang" saw (the "hog") —a series of upright saws, standing parallel with each other, having an up and down motion—was introduced. They could devour whole logs at one time.

A visitor to East Saginaw in 1868 wrote,[36] "within the circle of our vision we can count forty-eight chimneys and smokestacks indicating as many manufactories of lumber, salt,[37] shingles, staves, etc. The Saginaw River, stirred by the incessant turmoil of the busy tugs, flows steadily on in its northerly course. Great vessels and small ones lie at their docks discharging or receiving freights. The frequent screams of the tug whistles drown the busy clangor of the nearest mills, and silence the rumbling of the various street cars. The great draw in the city bridge swings at the summons of an approaching tug that pulls after it a heavily laded lake vessel," and then, "looking away to the horizon the sky seems to settle down to the fringe of forest."

By the early 1850's the Saginaw river was fairly alive with tugs, rafts, barges, lake schooners, packets, Mackinaw boats, bateaux, paddle-wheelers—anything that would float and carry, haul, pull or push logs, lumber, supplies, machinery or human beings. In one month alone in 1868, 1,694 tugs, 442 sailing sloops, 217 barges and 326 steamers passed through the Third Street bridge in Bay City.[38]

Almost every kind of craft sailed the rivers and lakes carrying lumber, shingles, laths, logs, salt, shooks, tanbark—and supplies for the mill towns and lumber camps. Above is the schooner "Day Spring—" at the left, the steamer, "Sidney O. Neff."

TEFFT'S
Saw Sharpener and Gummer.

PILE DRIVER "O. MASON."
MASON & STARK, Owners.

Are Prepared to Build

DOCKS & BOOMS,
AND DO

PILE DRIVING,

Anywhere in the Saginaw Valley, or Lake Huron Shore.

E. J. HARGRAVE & SON,
Manufacturers of

LUMBER AND LATH

Our Facilities for Sawing Lumber by Contract are unsurpassed in the Saginaw Valley.

OFFICE AND MILLS ON MIDDLE GROUND.

BAY CITY

Jackson, Lansing & Saginaw
RAIL  ROAD.

OPERATED BY THE

MICHIGAN CENTRAL R. R. CO.

THE MOST DIRECT ROUTE TO THE GREAT LUMBER & SALT REGIONS OF THE SAGINAW VALLEY, AND THE MAGNETIC MEDICINAL SPRINGS OF ST. LOUIS.

Jackson, Lansing, Owosso, Chesaning, St. Charles, Saginaw City, East Saginaw, Wenona, Standish, Wells, West Branch, St. Helens and Roscommon & Crawford on Main Sable River.

JESSE HOYT,
Gang Sawed Lumber.
LUMBER DELIVERED ON CARS FREE OF CHARGE.

Dressed Lumber a Specialty.

W. J. BARTOW, Agent.

OFFICE, BUENA VISTA BLOCK.

Mills and Planing Mill, East Saginaw, Mich.

ZILWAUKIE MILLS.
RUST, EATON & CO.,
Dealers in Gang-Sawed

PINE LUMBER,
Saginaw City, Mich.

40,000 ACRES
OF VALUABLE

PINE LANDS
FOR SALE.

These Lands are situated in the Counties of

SAGINAW, MIDLAND, BAY, GLADWIN OGEMAW, ROSCOMMON AND IOSCO, MICHIGAN,

ARE

WELL TIMBERED
And Eligibly Situated for Lumbering,

The Saginaw Valley Land, Salt & Mineral Company

JAMES STEWART & CO.,
Wholesale Dealers in

CORN, OATS, FEED
FLOUR, PORK,

AND

Lumbermen's Supplies Generally.

ALSO, DEALERS IN

KEROSENE, LARD & WEST VIRGINIA OILS, WATER LIME, FIRE AND CLAY BRICK, &C.

☞ *ALL GRADES OF SALT CONSTANTLY ON HAND IN CAR OR CARGO LOTS*

BUTMAN & RUST,
Manufacturers and Dealers in

GANG SAWED LUMBER,
AND DEALERS IN

LOGS & PINE LANDS,
Office, Bernhard's Block, SAGINAW CITY.

BANCROFT HOUSE.

ISRAEL B. NORCROSS, Prop'r,

EAST SAGINAW, MICH.

Wagons, Sleighs, Lumbermens Bobbs
CARRIAGE IRONING, MILL REPAIRING, GENERAL BLACKSMITHING, HORSE SHOEING, &C., DONE PROMPTLY.

I. A. SHANNON & BRO.,

BAY CITY, MICH.

The "Charles H. Bradley," with consorts "Brightie," "Mary Woolson" and "Goshawk" in tow.

A. W. Wright's wharves and mill pond—Saginaw City. Logs strung together with rope and rafting pins, await their fate in the quiet mill pond.

The noisy mill towns were wide open and rip-snorting—and just as wicked as the Reverend Henry Chance, otherwise known as "Buck-Eye Broad Axe," and other fire-breathing revival-temperance lecturers said they were* These towns, with their gaslights, saloons, sawdust and bawdy entertainment, were where the shanty boys and riverhogs, teamsters, ink-slingers, pin whackers, pushers, raftsmen and scalers always headed, looking for excitement. The riverhogs came from the river drives, and the shanty boys from the pine woods—cocky, full of beans and ready to wet their whistles, start a fight or see the girls, before they headed back (with empty pockets) to the camp, to another drive—or to a duller life on the farm or in the village.

*In 1876, there were 306 saloons in Saginaw—and in Bay City, Water Street was so notorious it was called "Hell's half mile."

Most everything in the mill towns was made of wood, and every business revolved around lumber. Fires were so common that in the '80's tugs and steamers were required by law to cover their smokestacks with "spark catchers" to keep flying, red hot sparks from igniting wood frame buildings, sawdust piles and the miles of drying lumber on river wharves. Every Spring, when the river rose and spilled over low-lying banks, citizens in Saginaw and Bay City were not surprised to see their plank sidewalks and cedar-block pavements float out to Lake Huron—and when the wind was in the right direction, lake sailors on incoming vessels could sniff the air and smell the sweet fragrance of pine resin and sawdust from the mills as their ships approached the mouth of the Saginaw from the bay.

Final Chapter

The logging industry in the Saginaw Valley reached its all time peak in 1882* when the once-unspoiled virgin forests of white pine were almost depleted. Many lumbermen wasted no time in abandoning the Saginaw for the still uncut forests in northern Michigan, Wisconsin, Minnesota and the Northwest. Others, in desperation, began to import logs from Canada which they towed in giant-sized "bag-booms," across two hundred and thirty-eight miles of treacherous Lake Huron from Georgian Bay, to keep their hungry mills operating and the ailing industry alive. But it was doomed to die, and in 1898 when Ontario decreed that "all logs cut on crown lands must be manufactured within Canada**,[39] log rafting stopped, almost over night. The

*Production on the Saginaw River in 1882 was over one billion board feet of lumber.

**In retaliation to the 1897 $2.00 import duty which the United States restored on white pine lumber.

sawmills began to move across the border. By 1897 the lumber production on the Saginaw River had reached the staggering figure of 22,930,757,551 board feet cut*.[40] Edward Cowles, statistician of the pine river trade in the Saginaw Valley, stated, "There were floated down the Saginaw River, in the life of the lumber harvest in the Valley, sufficient quantities of pine logs

*Not included in this incredible figure were the 5,580,535,223 shingles and other pieces that were cut. These figures represent lumber cut between the years 1851-1897.

to equal . . . the total national debt at the close of the Civil war."[41] When one stops to consider the loggers' rough estimate rule was "approximately 200 board feet to the log—or five logs to the thousand"—it is no wonder there was, throughout the entire industry, such profound respect for the little log mark. The log mark reigned supreme over all of the mighty Saginaw River Basin's logging operations—from remote pine forests to sawmill.

Today, there are few reminders left of the Saginaw Valley's once-gigantic logging and lumbering trade. Princely fortunes were amassed and lost—and in a few isolated instances, (perhaps to ease an uncomfortable conscience or two) several lumbermen left libraries, public buildings and parks, that still bear their names, to Valley communities. One can still see rotting spiles which protected river banks from floating logs, and others which mark the location of great wharves where lumber barges and schooners were loaded. There are also vanishing skeletons of old lumber hookers sunk in the mud along the Saginaw River banks—which recall to mind the long-gone days when they were loaded to their boom jaws with great cargoes of cut lumber, laths, barrels of salt, shooks, shingles and tan bark—and then proudly sailed across the lakes to eastern ports. The slowly

disappearing lumbermens' mansions (enormous dwellings with bulging bay windows, stained glass, broad porches and high cupolas) still occupy their positions of dignified eminence on old, tree-shaded streets in the sawmill towns. These are the survivals—the only visible reminders of the boisterous, greedy, extravagant, colorful days—when the cry "timberrrr" echoed throughout the Valley.

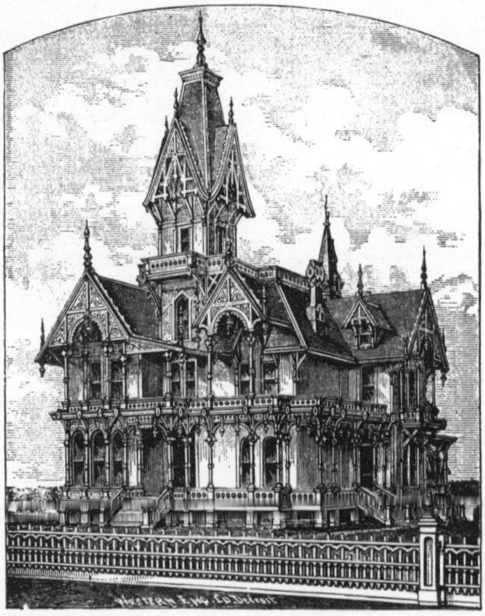

Logger's Lingo

AGUE—Malarial fever. Most early pioneers and loggers endured the "fever 'n' ager," caused by the bites of pesky mosquitoes that bred in swampy areas. Quinine was generally taken as a remedy.

AXE (branding)—see "Marking Hammer."

BOOM—An enclosure designed to hold back, encircle, or "fence in" floating logs.

BAG-BOOM—A giant-size raft made up of free-floating logs enclosed in a boom—designed to ride lake ground swells over great distances on the lake's surface.

BOOM-DOGS—Two 5½" iron wedges connected by four to eight (or more) links of chain. Wedges were driven into log ends, chaining them together to make a floating log "fence" or boom.

BOOMING GROUNDS—were located at the mouths of the main streams and were where the log drives terminated. Here, logs were sorted according to company log marks, gathered together into rafts with rope and hardwood pins, and towed to the sawmills. The largest boom company in the Saginaw Valley was the Tittabawassee River Boom Co. It owned over twelve miles of booms. During peak years, the Tittabawassee Boom Co. purchased 110 tons of rafting rope, which would "barely last 60 days." In 1872, the company spent $64,541 for rafting rope and almost $10,000 for rafting pins.[43]

BOOM-STICKS—a kind of narrow, wooden, plank catwalk, held up just above the water level by piling. Sorters stood on boom-sticks while sorting logs at the booming grounds.

BUCKING LOGS—Cutting or sawing felled trees into 12, 14, 16, 18 or 20 foot lengths. Using a cross-cut saw, two sawyers working together, could "buck" about one hundred saw logs a day.

CANT HOOK—A tapered wooden pole, three to six feet long with a seven to nine inch, loose, dangling hook near its working end. Cant hooks were used for loading logs on sleighs, on skids and for decking logs at the banking grounds or rollways. The cant hook man placed the hook on a log and bore down on the handle—gave it a pull and the log would cant over. A good cant hook man was highly respected in the logging camps. Canting logs was a dangerous job.

CHEAT STICK—The loggers' rule or scale—similar in appearance to a yardstick—with a metal tip on one end. It was used to measure logs, noting the number of board feet each contained. The two scales most commonly used were the

Scribner's Scale and Doyle's rule.

CHICADEE—The Chicadee's humble task was to keep the logging roads and trails free of horse manure.

CORKS—Also known as "caulks" and "korks." They were the long, needle-sharp spikes in the riverman's boot heels and soles—to keep him from slipping from the logs when he rode them down the swift streams.

CROSS-CUT—A saw—sometimes seven or eight feet long with cutting edge slightly curved or bow-like from end to end. It was equipped with a handle at each end so two men could work—one from each side of a tree, felling it and then bucking it into logs.

CROSS-HAUL—A sort of "way station" where logs were "snaked" or dragged by horses or oxen to await being transported to skidway or banking ground.

DEACON—The shanty boy who could sing ballads or recite verses —play the harmonica, "squeeze box," fiddle, guitar, dulcimer, jews-harp—or tell a good yarn. He usually performed for the men in the bunkhouse on Saturday night—sitting or standing on the "deacon's bench" while entertaining.

FLY BOOM—A boom or enclosure made of long timbers chained together, to be strung across a stream and fastened to a tree on the opposite bank whenever trouble threatened below. "Jam below!" brought immediate results.

GABRIEL—A long tin horn which "cookee" blew to call the shanty boys to their "hash." The gabriel sometimes was as long as eight feet!

GRAYBACK—Lice or bedbugs.

GUT HAMMER—A metal triangle on which cookee beat a tattoo to call the shanty boys to meals. (See "Gabriel.")

INK SLINGER—The clerk who kept time and production records in the logging camp. He sometimes doubled as a scaler.

JAM—When logs became "hung up" on sand bars, bends in river, on rocks or other obstacles, logs sometimes backed up on streams for miles. A log jam was one of the most dangerous and contant hazards of a river drive.

JOBBER'S SUNS—Kerosene torches used for night work in the woods.

KERF—The gash in a log, made by the saw.

LOG MARK—A brand or "mark," struck with a heavy marking hammer, on each end of all cut logs to designate ownership. The raised designs on the striking surface of the hammer were many and varied—triangles, circles, figures, and combinations of all of these. The log mark was first developed and used in Michigan, near Muskegon in 1842[44]—and was adopted throughout Michigan logging camps.

LOGGERS' SMALL-POX—Sawmill towns' wood plank sidewalks, saloon and dancehall floors—and the faces of victims of barroom brawls always showed signs of "loggers' small pox." It was a common malady during the brawling days of the logging era—marks left by the riverhogs' corked boots.

LUMBER HOOKER—A shallow-draft lake vessel that carried cargos of lumber to lake ports. In the early days of lumbering, the three-masted, fore-'n-aft schooners sailed the lakes, powered only by the wind—their holds filled and decks sometimes piled twelve feet high with lumber. The boom jaws were shifted up the mast to allow for the stacked cargo. Gradually, the windjammers were shorn of their sails and converted to barges. They were loaded with lumber and towed by tugs or wooden steamers. It was not unusual to see a "sow and her pigs"—a steamer or tug followed by a string of three, four or five consorts capable of transporting over 3,000,000 feet of lumber.[45] As years passed, wooden steamships grew longer and were strengthened to carry greater loads of lumber in their holds and on their decks. They became the forerunners of the steel freighters we see on the lakes today.

LUMBERJACK—(See "Shanty Boy").

MARKING HAMMER—Also called stamping ax, "iron," branding ax, stamp hammer and marking iron. A hammer-like tool that was used to strike the owner's mark on the ends of his logs. The striking surface of the hammer had the owner's mark raised on it and sharpened so it would easily cut into the soft wood of the log. The A. F. Bartlett Company, a foundry firm in Saginaw, was for long the chief maker of marking hammers in the Saginaw Valley.

PEAVEY—The riverman's tool invented by Joseph Peavey of Maine in 1858. A stout, tapered, wooden pole approximately five to six feet long, with a steel or iron point on its end and a large, dangling hook that opened to eight or ten inches. The peavey was used for turning, rolling, jabbing and hooking logs piled on top of each other—and to free them from obstacles or from the banks of the rivers.

PIKE POLE—A long wooden pole with a sharp iron or steel point and a small, spur-like stationary hook. Pike poles were from six to twenty feet long and were used by rivermen to jab, push and pull logs "hung up" in the streams—or at the booming grounds in the sorting pens.

PIN WHACKER—Usually a lightweight man or boy who worked in one of the many sorting pens or pocket-booms at the booming grounds. His job was to stand on each log as it floated into the pen, and with wooden mallet, drive a staple-like hardwood rafting pin over a length of rope, pinning it to the center of the log. As each log entered the enclosure, it was pinned, roped and drawn up next to its neighbor—thus a raft was made—ready to be towed to the sawmill.

PUSHER—The foreman or boss of the logging camp—sometimes called the "push" or "bull-of-the-woods."

RIVERHOG—Also called riverpig, white-water man, riverjack, riverdriver. These men were the agile, skillful, fearless individuals who drove the logs on the rivers from the rollways to the booming grounds.

ROAD-MONKEY—Road builder in the logging camp.

ROLLWAY—Also called a landing, banking or decking ground. An inclined way on the river bank, where logs were decked or stacked horizontally upon each other—or piled like giant jackstraws down the bank and on the frozen streams

—during the winter months. This was the starting point of the river drives.

'ROUND FORTY—A dishonest way to cut timber. A logging company, after purchasing a piece of land, would not only cut the timber on that piece, but dishonestly cut trees on land surrounding it. Legislation, making the practice illegal, was effective in 1903—but it was too late to benefit the State of Michigan. Most of the white pine had been cut.

SCALER—The man who estimated how many board feet of lumber there were in each log at skidway or rollway. The scaler used a loggers' rule ("cheatstick") to measure the logs.

SHANTY BOYS—Loggers, woodsmen. (Lumberjack or "jack" is of recent literary origin and was seldom used by the workers themselves[46]). These were the hardy men who lived and worked in the logging camps—chopping down and sawing the white pine trees into logs—swamping, skidding, and finally loading the logs on the sleighs and decking them high at the rollways.

SHOOKS—Barrel staves, hoops and heads—shipped knock-down to save space.

SKIDDING TONGS—Steel tongs, similar to huge ice tongs. They were clamped on one end of a log which was then dragged or "snaked" to a "cross haul" or skidway by a team of horses or oxen.

SKIDWAY—Platforms of logs piled at way-stations along the sleigh roads. The logs were loaded on sleds at these stations and hauled over iced roads to the rollways on the river banks.

SLEIGH—Also called bobsleigh, bobsled or sled. The major vehicle for transporting sawlogs from the skidways to the rollway. Runners on a sled were six to eight inches wide and about eight feet long. Fastened above each pair of runners was a ten to sixteen-inch-wide wooden cross beam known as a "bunk." The logs were loaded on the bunks and hauled by teams of horses over iced roads to the rollway. A loaded sleigh of six thousand feet of logs, or sixty thousand pounds was not an unusual one.

SORTER—Also called gap sorter. Sorters were stationed at the entrances (gaps) of individual enclosures or pens located off each side of the long central channel of the booming grounds. The sorters stood on catwalks (boom-sticks) and as the logs floated down the channel toward them, they looked for their company's log marks stamped on the log ends. The identified logs were poled with long pike-poles, into the pens where the accumulated logs—all bearing the same marks—were made into rafts and then towed to the mill.

SNUBBERS (Snubbing)—Sleigh roads were usually down inclined ways leading to the banking grounds. A common way to keep the heavily loaded sleigh from running over the horses (called "sluicing a team") and spilling the load, was for men to hook ropes to the sled, then loop the other ends of the ropes around two or more stout trees or stumps to hold the sleigh in check. The sleigh was then gently eased down the treacherous iced road by gradually let-

SNUBBERS—Continued

ting out on the taut ropes until the sled was on level ground once more. Any braking device on sleigh or wagon was called a "snubber."

SPRINKLER—A device used to ice the sleigh roads. A square, strongly-built wooden tank, holding as high as one hundred barrels of water, was placed upon a pair of sleds with a tongue at each end to save turning it around. The tank was generally built as wide as the log road, or the width of the sled, which had eleven to fourteen foot bunks. The water escaped from two holes located directly over the runner tracks. On cold nights the water froze almost instantly on reaching the road. Icing the sleigh road was done at night by teamsters ("icers") so as not to interfere with teams hauling logs by day. Crude boilers were sometimes used to keep the water in the tank from freezing before it reached the ground.

TEAMSTER—the man who drove the horses, mules or oxen—hauling supplies, driving the bobsled, skidding logs or icing the camp sleigh-roads. The teamster was usually hired along with his own team of horses.

TRAVOIS—sometimes called a "go devil,"—a short, heavy sled (many times cut from a forked branch of a tree)—used to support one end of a log or several logs that were dragged by a team over the ground to the skidway.

VALLEY BOYS—Saginaw Valley loggers had legendary reputations of being rough and tough—who could outfight, outdrink, outsaw any other "blinkity-blankity" shanty boy in the upper or lower peninsula of Michigan.

WANIGAN—also spelled wannegan, wangan, etc. The word had different meanings and spellings in different camps and in various parts of the country . . . but most always a wanigan was known as a scow or raft with a raised deck, and a pine slab shack, log cabin or canvas tent perched on top. The wanigan was kitchen, office, supply room, dining shack and camp store (van) during the spring river drives. Several wanigans were usually chained together and floated down the logging streams behind the rivermen to serve them while the drive was on.

WHITE PINE—A cone-bearing evergreen that reaches to a height of about eighty to one hundred fifty feet, and higher—the trunk measuring as much as four to seven feet in diameter. The needles of the white pine come in bundles of fives and are pleasantly aromatic. Its cones droop and are from three to five inches in length—sometimes maturing to ten inches. The Michigan White Pine was the most valuable to lumbermen of the three native pine trees (White, Red or Norway, and "Jack" Pine) because of its height, buoyancy, its straight trunk and its smooth almost white wood. The only stand of virgin white pine left in the lower peninsula of Michigan can be seen at Hartwick Pines State Park near Grayling. It is perhaps ironic that the white pine is the state tree of Michigan.

WIDOW-MAKER—A long limb on a tree which could fall on a man and kill him.

Notes

1. Rolland H. Maybee, "Michigan's White Pine Era—1840-1900." Pamphlet No. 1, John M. Munson History Fund Publication (Michigan Historical Commission—Lansing, 1960).
2. John Kieran, "An Introduction to Nature," 160 (Garden City, N. Y., 1955).
3. Mrs. Anna Brownell Jameson, "Winter Studies and Summer Rambles." Vol. 2, page 214. (Wiley & Putnam, New York, 1839).
4. George Newman Fuller, "Economic and Social Beginnings of Michigan." 51, (A story of the settlement of the lower peninsula during the territorial period—1805-1837), Lansing, Michigan, 1916.
5. Frank B. Woodford, "Lewis Cass—The Last Jeffersonian," 108, (Rutgers University Press, New Brunswick, N. J., 1950).
6. American State Papers: Public Lands, III: 164-165.
7. Frank B. Woodford, "Lewis Cass—The Last Jeffersonian," 108.
8. Ibid., 113.
9. J. Harold Stevens, "The Influence of New England in Michigan." From Vol. 19:329, Michigan History Magazine.
10. H. M. Utley, "Wildcat Banking System." From Vol. 5:209-211 Michigan History Magazine.
11. John T. Blois, "Gazetteer of the State of Michigan." (Detroit and New York, 1838) Under "Saginaw County."
12. Ibid., under "Lower Saginaw."
13. Philip P. Mason, "The Lumbering Era in Michigan History 1860-1900," (manual accompanying film strip) 5. Prepared for Michigan Historical Commission through the Munson Michigan History Fund.
14. J. Harold Stevens, "The Influence of New England in Michigan." From Vol. 19:328, Michigan History Magazine.
15. George H. Hazelton, "Reminiscences of Seventeen Year Residence in Michigan, 1836-1853." From Vol. 21:370-417 Michigan Pioneer and Historical Collections.
16. William H. Hathaway, "County Organization in Michigan." From Vol. 2:573, Michigan History Magazine. Government land offices were established in Detroit, 1818; in Monroe, 1823; Kalamazoo, 1831; Flint, 1836; Ionia, 1836.
17. Rolland H. Maybee, "David Ward: Pioneer Timber King." From Vol. 32:1-14, Michigan History Magazine.
18. William G. Rector, "Log Transportation in the Lake States Lumber Industry." 46 (Glendale, California, 1953).

19. Ibid., 48.

20. Lewis C. Reimann, "When Pine Was King." 9 (Ann Arbor, 1952). Charles Moore, "History of Michigan." Vol. I-25:513 (New York, 1915).

21. Landon, "Lake Huron." 104-105, (The American Lakes Series, M. M. Quaife, Editor) Detroit, 1947. W. L. Clements wrote, "The great period of Michigan's lumbering days, may have been from 1862 to 1880. During these years was begun and carried to consummation, without regard to conservation or the rights of future generations, the destruction of Michigan's white pine forests. Saginaw Valley was the focal center of Eastern Michigan in this destruction; Saginaw River, the confluent stream through which logs passed to the mills. Along the river, mills were then in course of erection, commerce and population were increasing, which a little later were to make Saginaw and Bay City the busiest communities in the state, and to put an accelerated activity into twenty years, which with today's advanced ideas of conservation, should have been spread over a century."

22. Ormond S. Danford, "Social and Economic Effects of Lumbering on Michigan—1835-1890." From Vol. 26:352 Michigan History Magazine.

23. Anita Shafer Goodstein, "Biography of a Businessman—Henry W. Sage, 1814-1897." 73 (Cornell University Press, Ithaca, New York).

24. Maurice Edron McGaugh, "The Settlement of the Saginaw Basin." 157 (Chicago University Press, 1950).

25. William G. Rector, "Log Transportation in the Lake States Lumber Industry." 17-20.

26. Philip P. Mason, "The Lumbering Era in Michigan History 1860-1900." Page 10 of manual. "As logging increased in the 1840's and 1850's a complex problem arose regarding the identification of logs. In 1842 and 1859, the Michigan legislature passed laws requiring owners to mark ends of logs and to register them in the county in which they were to be sawed."

27. William G. Rector, "Log Transportation in the Lake States Lumber Industry." 112.

28. Ibid.

29. Ibid., 226.

30. Maurice Edron McGaugh, "The Settlement of the Saginaw Basin." 38 and 128.

31. James Cooke Mills, "History of Saginaw County, Michigan." 393-404, (Saginaw, Michigan, 1918).

32. Maurice Edron McGaugh, "The Settlement of the Saginaw Basin." 164.

33. Lumberman's Gazette, XI, 100.
34. Bay City Directory, 1868. 264.
35. James Cooke Mills, "History of Saginaw County." 393-404. Anita Shafer Goodstein, "Biography of a Businessman." 28. The first shipment of clear cork pine was sent by Curt Emerson from Saginaw via the Erie Canal to Albany, N. Y. in 1847. It was consigned to C. P. Williams & Co.—the firm in which Henry W. Sage had an interest. Fifteen years later, Sage went west and built one of the largest sawmills in the world in West Bay City (which he founded and called Wenona) and made a fortune in the manufacture of lumber. (Authors' note)
36. East Saginaw Directory, 1868-69. Page 3.
37. William L. Webber, "Discovery and Development of the Salt Interest in the Saginaw Valley." From Vol. 4:13, Michigan Pioneer and Historical Collections. Salt was a by-product of the lumbering industry. Test wells were sunk in 1839 on the Tittabawassee River at the mouth of the Salt in Midland County—but it wasn't until there was cheap fuel (wood slabs, sawdust, refuse) available from the sawmills to provide heat for the evaporation of the brine, that salt production became big business in the Valley. The first commercial well was sunk in East Saginaw in 1859. (Authors' note)
38. Bay City Directory, 1868-69. Page 75.
39. Robert C. Johnson, "Logs for Saginaw: An Episode in Canadian/American Tariff Relations." From Vol. 34:213-223 Michigan History Magazine.
40. James Cooke Mills, "History of Saginaw County." 403.
41. J. A. Russel, "Story of Michigan's Marketing." From Vol. 13:33 Michigan History Magazine.
42. William G. Rector, "Log Transportation in the Lake States Lumber Industry." 167-169.
43. Ibid., 135.
44. Clifford Allen, Editor, "Michigan Log Marks." 11. A WPA Writers' Project (Michigan State College Memoir Bulletin—No. 4, East Lansing, 1942).
45. Walter Havighurst, "The Long Ships Passing," 119. (Macmillan Co., New York, 1942).
46. Rolland Maybee, "Michigan's White Pine Era—1840-1900." 27. Pamphlet No. 1, John M. Munson, Michigan History Fund Publication, Lansing, 1960.

We wish to acknowledge with gratitude the following people who assisted us in our research for the material used as a basis for the production of this little volume. Their able assistance contributed immeasurably to the final result.

Mrs. A. T. Greenman, Curator, and Miss Zella Entrekin of the Bay County Historical Museum; Mr. Ralph Stroebel of Saginaw, for his log mark research; Miss Jennie Kulberg, Mrs. Eugene FitzGerald and Mrs. Budd Graveline of the Sage Public Library in Bay City; Miss Ann Ballou of the Bay City Public Library in Bay City; Mrs. Nelda Hinz, Miss Gladys Blakely, Miss Margaret Bray and Mr. Monroe Causley, Jr. of the Hoyt Public Library in Saginaw; the staff at the Butman-Fish Public Library in Saginaw; Mrs. Alice Dalligan and members of the staff of the Detroit Public Library; Miss Esther Laughin of the Michigan State Library in Lansing; Captain H. C. Inches, Director of the Great Lakes Historical Society Marine Museum, Vermilion, Ohio. We want to especially give our thanks to Miss Geneva Kebler, Reference Archivist, and her co-worker Mrs. Elizabeth Rademacher of the Michigan Historical Commission Archives in Lansing for their courtesy and time spent in reading over our manuscript and for their helpful comments. To these people we give our thanks.

The Authors

The photographic illustrations appearing in this book are either photographs taken by the authors—or were photographically reproduced by the authors from the following collections.

Courtesy, MICHIGAN HISTORICAL COMMISSION ARCHIVES:
Cover, inside front cover, title page; pages V, 4, 12, 15, 16, 18, 20, 21 (lower), 22, 23, 24, 25, 26, 27, 28, 29 (top), 30, 32, 33, 34, 35, 41 (top), 44 (top), 45, 47, 50, 51 and 54.

Courtesy, BAY COUNTY HISTORICAL MUSEUM: Pages 14, 21 (top), 29 (lower) and 44 (lower).

Courtesy, HOYT PUBLIC LIBRARY, Saginaw: Pages 46 (top), 49 (top and bottom).

Courtesy GREAT LAKES HISTORICAL SOCIETY MARINE MUSEUM, Vermilion, Ohio—Captain H. C. Inches: Pages 46 (lower) and 49 (center).

AUTHORS' PHOTOGRAPHS: Page 5—Gazetteer from the Burton Historical Collection, Detroit Public Library; pages 10, 31, 36 and 41 (lower) are items in the Bay County Historical Museum collection; 37, 43, 48, 55 and inside back cover.

Index

Watch her, catch her,
jump up on her jub-er ju . . .
Give her the sheet and let her slide,
The boys will push her through.
You ought to seen us howling,
The winds were blowing free,
On our passage down to Buf-fa-lo
from Bay Ci-ty.

A relic of the Saginaw's past—sunk in the mud along the riverbank.

KO
L
DAVE
B38
S
BUB
US
AN
FLY
HO
YT
C
BUB
JBW
IKE
BU
RT
MITT
DA
H A
L L
S.M
J
JWL
FREE
JAKE
JO
M
KING
HALL
CER
CL
PERL
W
JBF
MOP
5 2
01
13
I
LY
329
8
NOR
WAY
BUT
DEB
W
4R4
AX
$XMC
MIKE
PETER
D
3
8
DAVE
B
KIP
%A
H&L